# The Cairnryan Military Railway
## Bill Gill

Cairnryan looking north in the 1950s,

© Bill Gill, 2021
First published in the United Kingdom, 1999,
as The Cairnryan Military Railway 1941-1959
by Stranraer and District Local History Trust
This edition published 2021
by The Oakwood Press
Oakwood Series X103
54-58 Mill Square, Catrine, KA5 6RD
01290 551122
www.stenlake.co.uk
ISBN 978-0-85361-762-4

The publishers regret that they cannot supply copies of any pictures featured in this book.

Printed by
P2D Books, 1 Newlands Rd,
Westoning, Bedford MK45 5LD

Bill Gill died shortly after completing the manuscript for this booklet in 1999. In view of his contribution to the study of local history as a teacher, lecturer and researcher, it seems appropriate that this book is dedicated to him.

# Further Reading

The following were the principal books used by author during his research. None are available from the Oakwood Press; please contact your local bookshop, reference library or search for them on the internet.

Christensen, Mike, "Cairnryan Military Railway", *British Railway Journal*, 1995
Holme, Richard, *Cairnrvan Military Port 1940-1996*, 1997
MacHaffie, Fraser G., *The Short Sea Route*, 1975
Meacher, Charles, *Steam Sapper*, 1992
Murchie, A. T., *The Mulberry Harbour Project in Wigtownshire 1942-1946*, 1993
Murphy, Major S. P., "Military Ports Nos 1 and 2", *Royal Engineers Journal*, 1993
Smith, David L., *Legends of the Glasgow and South Western Railway in LMS Days*, 1980
Smith, David L., *The Little Railways of South-West Scotland*, 1969
Smith, David L., "Rusting Rails by Lochryan", *Railway Magazine*, 1971
Starling, Major J., *Pioneer Service at No. 2 Military Port*, Unpublished MS
White, Brigadier B. G., "Construction and Operation of Military Ports in Gareloch and Cairnryan", *Royal Engineers Journal*, 1958

# Acknowledgements

Line drawings from author's originals by Mr Peter Holmes. Ordnance Survey maps, Scale 1:10,560, Sheets NX 06 NE and SE, 1957 and 1967. The redrawn maps are reproduced with the permission of the National Library of Scotland. Photographs from The Imperial War Museum photographic archive; Stranraer Museum (The Bill Gill Collection) and Mr Donnie Nelson. Reminiscences from Mr Charles Collins, Miss Dalrymple-Hamilton, Mr George Hill, Mr Harry MacKenzie, Mr Gordon Muir, Mr Donnie Nelson, Mr Ian Shaw, Mr Alan Wallace, and the late Mr Gilbert Kelly, Mrs McCracken, and Sam Thomson.

The publishers would like to thank, Julia Macdonald of Stranraer and District Local History Trust for her help with this book, Norfolk Heritage Steam Railway and particularly David Bramhall for the information on Austerity No. 3193. And to Billy McCrorie for the use of photographs that he had uploaded to the Geograph website, showing some of the remains of the railway.

Sites considered for military ports in 1940.

# Introduction

Britain at war was utterly dependent on her ports. Through them passed, as well as commercial imports and exports, military reinforcements, arms, and ammunition for garrisons and theatres of war overseas. Commercial and military traffic did not mix well. Priorities were different, bottlenecks frequent, tempers short. The Dunkirk evacuation and the subsequent fall of France had by the end of June, 1940 brought about a direct threat to the ports of the South and East coasts of England now within Luftwaffe bomber range. With the large ports in the west facing the same threat at an early date, it was decided to solve the two problems of incompatibility and possible destruction as speedily as possible by the construction of two new ports, built by the military for the military.

Major S. P. Murphy RE in *The Royal Engineers Journal*, 1993, tells how the Director General of Transportation and Movements authorised, in July, a reconnaissance of the West Coast "to investigate the possibility of the construction of a new military port on the West Coast of Scotland with the object of freeing existing port facilities now in use."

A new port would not be part of an existing port. Other factors to be taken into consideration would be the availability of land for the housing of personnel, easy road access, adequate anchorage facilities, and good railway communication. Speed of construction would be essential and economies would be observed.

Major Murphy relates that in six days in the third week in July the reconnaissance party visited Barrow, Maryport, Stranraer, Ayr, Troon, Irvine, Ardrossan, Gareloch, Craigendoran, and Oban, coming to the conclusion that the sites at Gareloch and Stranraer were possible locations.

Strangely, before 1940 the British Army had very little experience of operating ports and even less of constructing them. A panel of eminent civil engineers gave its approval to the detailed surveys by the middle of September. Ideas were produced that could be put into immediate practice. American experts in civilian clothes visited the sites and a

A mounted detachment of Royal Engineers serch the moors above Craigammin Ridge for water sources and possible campsites.
*Nelson-MacDonald Collection*

detachment of Royal Engineers, on horseback, searched the moors above Cairnryan for water supplies and potential campsites.

Once the new ports were completed there would be no problems arising from commercial usage, as the ports would handle military shipping only. Haunted by memories of "Red Clydeside" in the First World War, work would now go ahead in military fashion. There would be no strikes over pay and conditions. Hours would be long with no breaks for inclement weather and living conditions would be absolutely basic at first and only slightly better afterwards. No civilian could be expected to work under those conditions.

A transcript of a talk by Brigadier B. G. White, the Director of Ports and Inland Water Transport, 1940-45, to the Institution of Engineers and Shipbuilders in Scotland in 1948 describes an amusing incident. He, the Brigadier, was given the unenviable task of informing the Minister of Labour about the decision of the War Office to employ only military labour on the construction of the ports. The Minister was Ernest Bevin, Secretary and founder of the Transport and General Workers' Union, the dockers' champion and a very formidable character. To the Brigadier's amazement and relief, Bevin agreed wholeheartedly with the plan and indeed became so enthusiastic that he insisted that the ports should be built large enough to accommodate the *Queen Mary* and the *Queen Elizabeth*. The War Office tactfully allowed this suggestion to founder on a sea of cost.

Faslane in the Gareloch was held to be the more advantageous of the projects and received priority from the start. Cairnryan had to fight off the claims of its two rivals, Stranraer and Wig Bay. The latter was the more sheltered site but to bring it up to deep sea standard would require much dredging that would be time-consuming and costly. Stranraer had rail and road connections but as a ferry port it had to be disqualified under the original stipulation. Though open to south west gales, Cairnryan had been noted for some two hundred years for its deep water facility.

David L. Smith, the first writer to generate an interest in the Cairnryan Railway, in an article in the *Railway Magazine* (1971) tells how Lord Inchcape was prepared to move his P&O base from Southampton to Cairnryan if war threatened. Dredging was needed but would be minimal, land for camp sites was available, and it was serviced by road from north and south. There was no railway but one could be built. The completion date would be August, 1942 for the whole project.

Priorities were set. The first would be a lighterage; the second would be the railway, on which would depend much of the construction of Military Port No. 2. Over seven miles in length, it was to be completed in eighteen months by Royal Engineers learning the job as they progressed, always short of skilled men and often of machinery and materials. They had the invaluable assistance of the Pioneer Corps, who swarmed like ants over the separate stretches in all sorts of weather, digging, draining, building, and hoisting. They were among the first troops there and a Pioneer company was one of the last to leave in 1959. This tribute does in no way detract from the magnificent work of the Engineers.

At first few problems presented themselves. For most of the way the track would be built on the level just above the High Water Mark (HWM). Work would begin from both ends simultaneously. The LMS main line, Dumfries to Stranraer, would feed in the rail traffic about two miles east of the latter. This was a single line but recompense for expensive improvements was promised from the Government.

An advance party of 29th Railway Survey Company RE arrived in Stranraer in November, 1940 to plan the route on the ground. Among their number was a young draughtsman, George Hill, who vividly remembers setting up a drawing office in the Stranraer Golf Clubhouse. He also recalls pitching a bell tent (in November!) on the first green. He further recollects that nobody had thought fit to inform the greenkeeper of their intention.

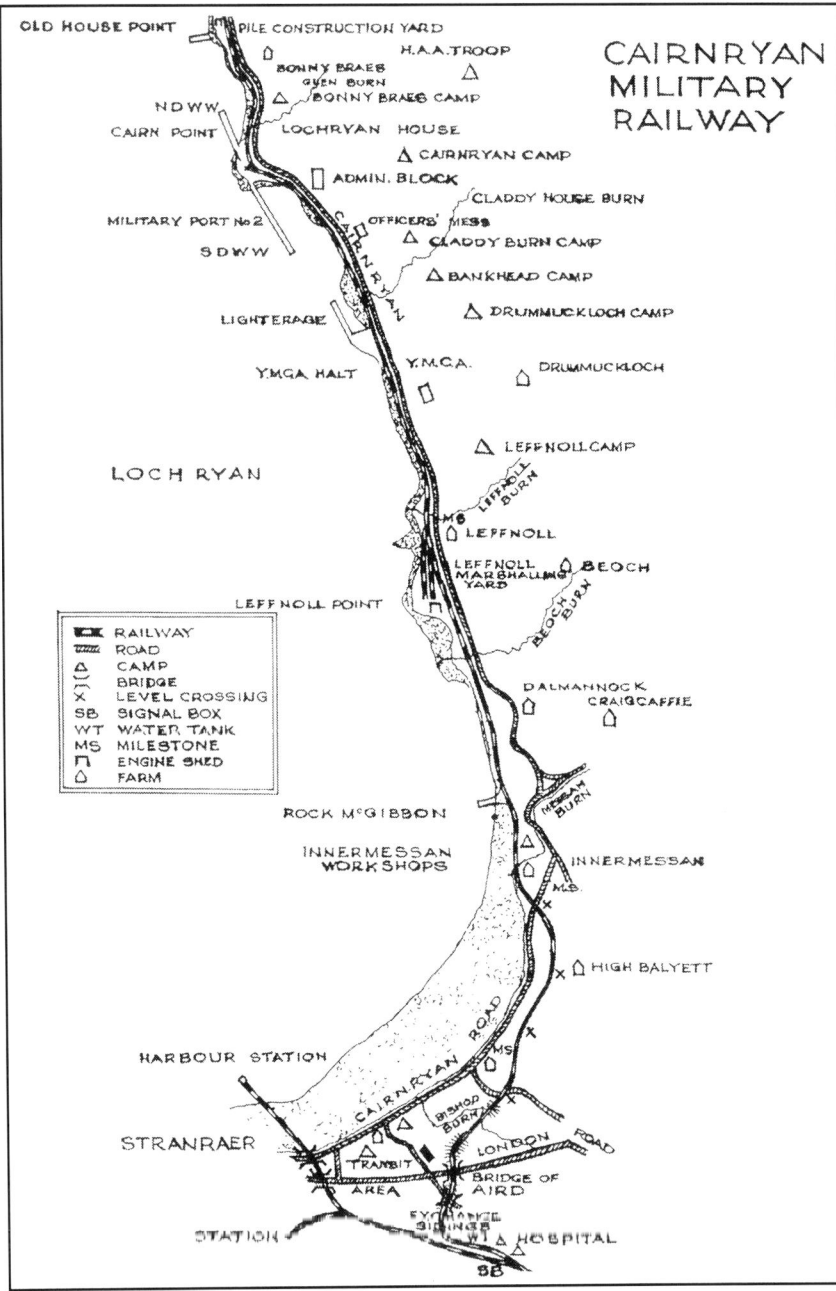

The ground plan, drawn up quickly, showed the military track leaving the main line at Castle Kennedy, curving round the outside edge of Lochinch Castle grounds and then heading directly to the shore at Innermessan. This plan, however, did not find favour with the War Ministry and a revised plan was demanded, which placed the rail junction west of the rail bridge on Limekiln Road, some fifty yards north west of the present kennels, about a mile from Stranraer.

One disadvantage of this plan, which was accepted, was the considerable down gradient from Aird to the Loch Ryan shoreline. Two major bridges would have to be constructed, linked by an embankment of some twelve hundred yards, sloping gradually and gently to the shoreline. The first bridge on the embankment would carry the line over the Limekiln Road by Bridge of Aird Farm. The second would straddle London Road. In between the bridges the Black Stank would have to be culverted. All this meant that the completion date could no longer be met, and therefore a temporary line would have to be laid around the obstacles and extended along the planned route. A small bridge over the Black Stank would take the new line to London Road, which it would traverse by level crossing. It would curve to the north east through what is now part of the Academy playing field, cross McMaster's Road and then follow the planned route along the flat ground. (In 1942 this temporary line was superseded by another, running along the eastern boundary of Aird Donald caravan park. This enabled the work on the embankment and bridges to be carried on from both ends.)

The new plan was approved and in January, 1941 the first of the RE construction detachments arrived at Aird, about forty strong. Bulldozers, scrapers, and other earth-moving machines arrived as well to establish a goods

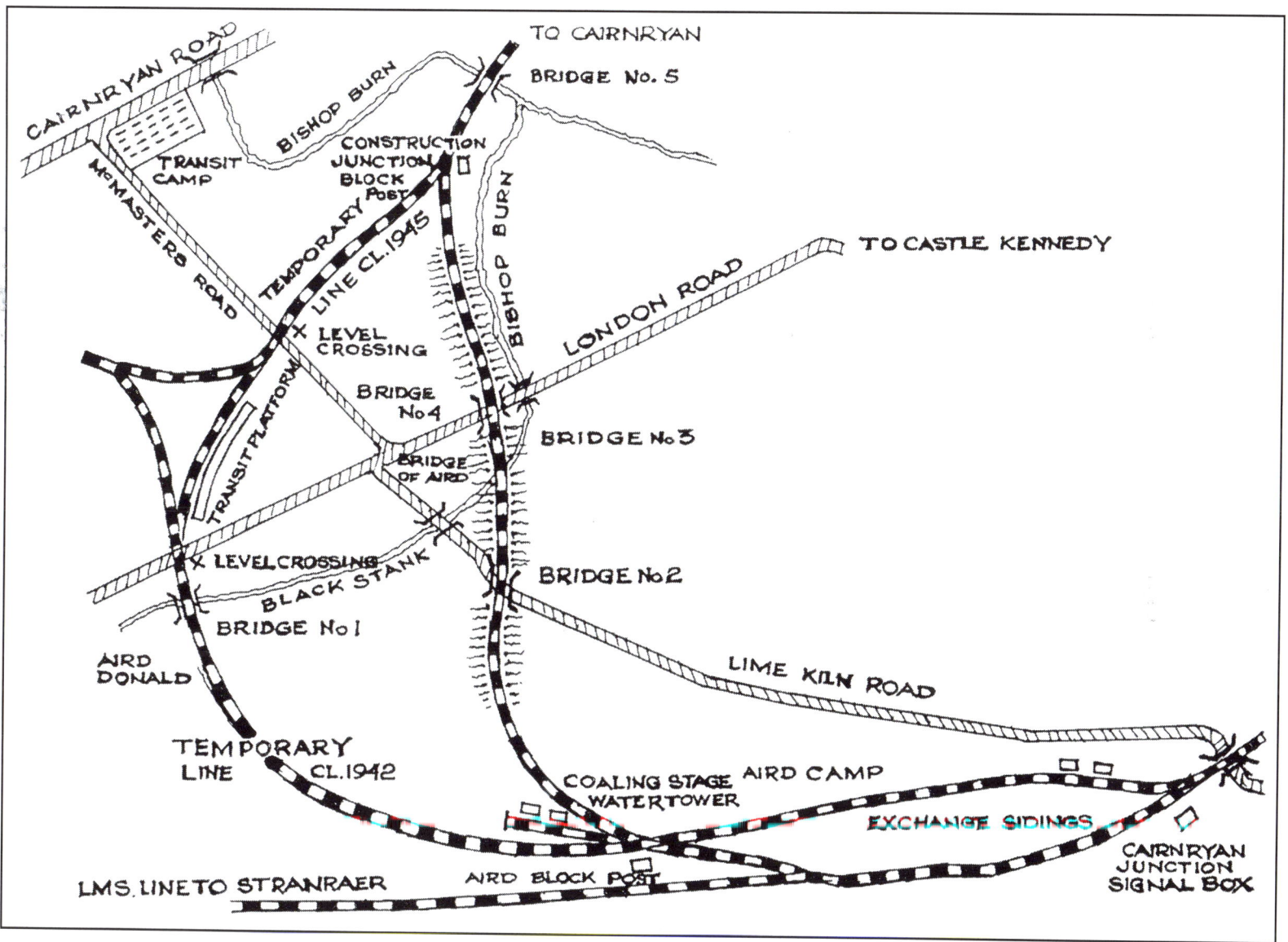

The start of the work tto create the exchange sidings at Bridge of Aird Farm, January 1941. Accomodation coaches, locomotive *Loch Moy* and the Stranraer line are on the left.
*Courtesy of The Imperial War Museum*

yard on the farmlands of Bridge of Aird. When this yard was levelled and graded it would allow the marshalling of hundreds of wagons from the LMS main line. It was designated Exchange Sidings. The weather was very cold, very wet, and very windy. No billets or huts were available for the workforce so the War Department, in a rare fit of compassion, provided five railway sleeping coaches and a recreational coach to be sited on a small siding off the main line. Welfare went even further with the arrival of railway locomotive *Loch Moy* to provide heating.

This was luxury compared to the bell tent pitched on Cairnryan Bowling Green by the small advance party at the north end of the projected line. Their office was the Bowling Pavilion. Their task was to prepare the way for another yard at Old House Point about half a mile north of the village. Here the manufacture of reinforced concrete for the piles and decking for the South Deep Water Quay was to be carried on. The yard was to share quarters with Cairnryan's only industry, the Lochryan Oyster Tanks, which somewhat miraculously survived the war. A wooden jetty was constructed at the north end served by a rail line. The pile driver was operated by a young sapper, Charles Meacher, whose book *Steam Sapper* has left us with an interesting and humorous account of life on the "Line". With the only problem the bridging of the Glen Burn, a line to the lighthouse area was soon in place and a temporary line was laid to the Lighterage site.

At the south end the temporary line extension east of Construction Junction had now become permanent and bridged the Bishop Burn south of Sandmill, crossing the minor road and the next two farm roads to Low and High Balyett by level crossings. Reaching the A77 at Innermessan, where a pole crossing was in operation, it now headed for the shore, curving round behind Innermessan Croft to bridge the Messan Burn alongside the old road bridge, at the foot of the 12th century mote. The speed of construction slowed down temporarily after this, reaching a point where tidal action had eroded the old road built at the end of the 18th century. An embankment now had to be constructed on reclaimed land protected from wind-lashed waves by steel plating held in place by iron stanchions.

Innermessan, or Rock McGibbon, Camp was to be built around the remnants of an industrial site dating from 1917, when the Multilocular Shipbuilding Company leased land from Stair Estates to build concrete ships in an effort to compensate speedily for the loss of Merchant Navy shipping caused by the unrestricted German U-boat campaign. Efforts to find out something more about this company have yielded little information. Charles Collins recalled that the manager, a Mr Dicks, became a businessman well ahead of his time in Stranraer. George Hill, having relinquished his share of a tent on the golf course, took himself to Innermessan Shore, where, in a First World War two-storey building he found, in a drawer of a draughtsman's table, a chart of Portsmouth.

The Innermessan camp was to be built to service the tugs and lighters working on what was to become the South Deep Water Wharf. It would eventually be staffed by RAOC and REME personnel. A slipway had survived from the First World War and a new one was added and the jetty renewed. A few sidings were built and the railway snaked through.

After leaving the campsite the line climbed slowly from the HWM, the embankment being protected for some yards by metal plates and then by boulders over a rubble core. It was given a concrete coating when it had passed a railwayman's hut, and picked out in pebbles on the concrete was a Pioneer Corps badge and date. In two hundred yards or so the line swung away from the beach and crossed the Beoch Burn on a small girder bridge. Later, Leffnoll Signal Box South would be built here. The line now became double track, at the entrance to Leffnoll Yard, a huge triangular marshalling area of forty-two acres with its base on the main road and its apex on Leffnoll Point.

Cairnryan Bowling Club *circa* 1920. The green became the site of the advance party's tent and they used the pavilion as their office. The gates of Lochryan House are on the far right of the photograph. Both they and the bowling club would be cleared to make way for the port facilities. In the background is Lochryan Church.

*Oakwood Press collection*

The YMCA building gifted by McVitie and Price, the Glasgow based biscuit manufacturers in 1943. It served as a canteen for the workers at Leffnoll marshalling yard. Later it became the Rhins of Galloway Hotel which is still providing accommodation to ferry passengers as it was in the 1980s when this photograph was taken.

*Nelson-MacDonald Collection*

To introduce a personal note, after the line became operational, a driver and his mate much enjoyed hospitality from two local residents, Mr and Mrs McCracken. Mr McCracken, who was the dairyman at Beoch Farm, became severely ill with pneumonia and his doctor, in the days before the discovery of penicillin, told Mrs McCracken that the temperature of the house must be kept as high as possible. Coal, of course, was rationed and the lady spoke of her worries to the soldiers. The driver told her not to worry but to "send the wee lad with a potato sack every morning to the Beoch Burn Bridge." The instructions were followed and the "wee lad" had the pleasant task of carrying a large lump or two of coal back to the house, Mr McCracken recovered.

As most of the traffic for the Military Port was at first planned to be outgoing, space had to be found for wagons waiting to be called forward for unloading at Lighterage and Rubble Bank, and empty wagons waiting to return to Exchange Sidings. Leffnoll was created to accommodate two thousand wagons. A very informative photograph from the Imperial War Museum collection (H40121) and an equally instructive diagram both feature in Mike Christensen's excellent articles in the *British Railway Journal*, numbers 54 and 55. On the diagram five groups of sidings are shown, amounting to forty-seven tracks. Buildings included a huge engine shed, erected in 1942, and, next to the main road, a large coaling stage. The through line ran parallel to the A77 until it passed Leffnoll North Box where a girder bridge took it over the Leffnoll Burn.

In 1943 a large YMCA building, gifted by McVitie and Price, the biscuit manufacturers, was opened by the Duchess of Gloucester. It had its own railway station directly opposite on the west side of the A77, a hundred yards north of the road bridge over the Leffnoll Burn. The station consisted of a wooden construction with a shelter.

At the north end of the line again, progress was being made, at a cost. The main road to Girvan curved left on leaving the

A train passing the YMCA buildings.
*Nelson-MacDonald Collection*

village, following the shore and passing behind the lighthouse and in front of Lochryan Parish Church. The proposed layout of the Military Port could not have a public road passing through it. Pressure was brought to bear on Wigtown County Council to engineer a new stretch of road branching north and running along the inner wall of Lochryan House grounds, bridging the Glen Burn some twenty yards back from the existing bridge. This gave the War Department the security it demanded and more than doubled the area at its disposal.

To make security absolutely certain, a wire mesh fence, six feet high, was erected from the church, running through the village (cutting the villagers off from the shore) and continuing for a further two miles to the Beach Burn road bridge. Lochryan House surrendered its deer park and its gate lodge. The village of course lost its bowling green. A relic of the past also

Cairnryan looking north. The six feet high fence cutting the village from the shore stands guard on the left.

*Nelson-MacDonald Collection*

A 1926 view of Cairnryan village looking south before the railway was built.

*Oakwood Press collection*

A similar early 1950s view of Cairnryan village looking south. On the left is the Lighthouse Block Post which was part of the signalling system on the line.

*Oakwood Press collection*

A *circa* 1910 view of Cairn Point from Bonybraes. The Lochryan Church (demolished in 1990) and the lighthouse are recognisable landmarks. Between them is the gate lodge and deer park of Lochryan House. Towards the shore from the Gate Lodge is a square of stone walls which is the bowling club.

*Oakwood Press collection*

A similar view of Cairn Point after the construction of the railway and port. The deer park, along with the gate lodge and bowling club have been cleared for railway sidings. Behind the church are the roofs of the North Transit shed. On the left are the massed huts of Bonnybraes Camp. A new road cuts inland from the church, through the old deer park. *Oakwood Press collection*

The gates and ornate gate lodge of Lochryan House *c.* 1910. In the distance on the avenue leading to the gates Lochryan House can be glimpsed. The wall behind the one that fronts the road marks the edge of the deer park, and became the boundary of the estate when the railway was built.

*Oakwood Press collection*

disappeared with the destruction of "the Pulpit", identified by C. H. Dick in his Highways and Byways in Galloway and Carrick, published in 1916, as "a round tower at the other end (i.e. south) of the wall."

Here before the erection of the village church the minister from Inch would come occasionally and preach to the congregation seated on the grass below. The surrendered land became part of another large marshalling yard, Rubble Bank.

Meanwhile, the second last part of the line had been completed, but not without difficulty. From Drummuckloch Road End to the Lighterage a high bank of the raised beach loomed over the main road, and the road ran right along the HWM. There was no space to site a single track much less a double one. Two choices faced the Engineers : build out into the Loch at expenditure of money and time, or divert the road. Given the situation the latter choice was a non-starter. The road could not be diverted but the high bank could and earthmoving equipment shifted a large part of it across the road to form the base of a new double track, level with or slightly above the road. The ground cut away below Bankhead, once levelled, became a car park.

The new embankment was protected by a rubble bank over infill. Storms that winter washed the infill through the rubble necessitating its replacement by a core of sand and gravel covered with an eighteen-inch layer of six to eight inch stone. A concrete "toe" protected the base of the construction and was carried back to the Leffnoll Burn estuary where scratched in the concrete was a record of the repair by 50th PC in 1943. The line now ran on in safety to the last stage.

Construction of the Lighterage at the south end of the village must have been well under way when the railway arrived as it constituted priority number one. It was built to shelter the tugs and lighters plying a steady trade across the developing South Deep Water Wharf and also to ferry cargoes from ships anchored offshore. The Lighterage was a completely artificial peninsula started in January, 1941 and composed of dredged

Mark of the 50th Pioneer Corps of their repair in 1943, at Leffnoll.
*Billy McCrorie*

material from the Loch and the contents of a quarry opened up for the purpose behind Lochryanhall. Named "Barrow" or "Borrow" Pit, the quarry served as a military transport park. A small marshalling yard was constructed and later a huge transit building was erected. The temporary single line from Rubble Bank was now replaced by a double track.

The small yard was named School Sidings, somewhat ironically as the arrival of the railway and the erection of the security fence turned out to be the death sentence for the village school. The building and the schoolhouse were demolished. About the same time and for the same reason, i.e. being on the wrong side of the fence, a clean sweep was made of Low Claddy House Farm, the fishermen's store, and the coastguard station and boathouse.

School Sidings was also called Claddyburn Sidings as here the Claddyburn was culverted for most of its length between the road bridge and the shore to make space for the sidings.

The very last lap lay between School Sidings and Rubble Bank. It was not without difficulty. The same problem had been encountered on the Drummuckloch to Bankhead stretch: there was going to be no room for the railtrack. The main street of the village ran most of the way along the HWM. It was narrow and housing stretched almost without interruption halfway along its length. The County Council again was pressurised into altering the road and the double track ran without hindrance to Rubble Bank by the lighthouse. The latter was one of the two buildings left standing on that side of the road, the other being the roadman's cottage or Strandmain Cottage by the entrance to the Lighterage.

Rubble Bank is sometimes referred to as "Transit" (which is somewhat confusing) because of the rail platform and the Transit Shed built there. This was the large area referred to earlier, made by the sacrifice of the deer park and bowling green, the last of the marshalling yards, with the appropriate number of sidings serving the South Deep Water Wharf and the later built North Deep Water Wharf. Here the double track ended as it joined the single-track line to the Pile Construction Yard. After eighteen months hard slog, the Line was completed. For some months, though, it had been partly operational before its official opening in July, 1942.

The late David L. Smith in his article already mentioned unearthed the following timetable information:

### May 2nd 1942

| so | so | sx | | so | so | sx |
|---|---|---|---|---|---|---|
| 13.15 | 17.15 | 18.51 | RUBBLE BANK | 21.36 | 23.06 | 22.49 |
| 13.21 | 17.21 | 18.57 | LEFFNOLL NORTH | 21.34 | 23.04 | 22.47 |
| 13.28 | 17.28 | 19.04 | INNERMESSAN | 21.27 | 22.57 | 22.40 |
| 13.37 | 17.37 | 19.13 | TRANSIT (STRANRAER) | 21.15 | 22.45 | 22.30 |

so – Saturdays Only; sx – Saturdays Excepted

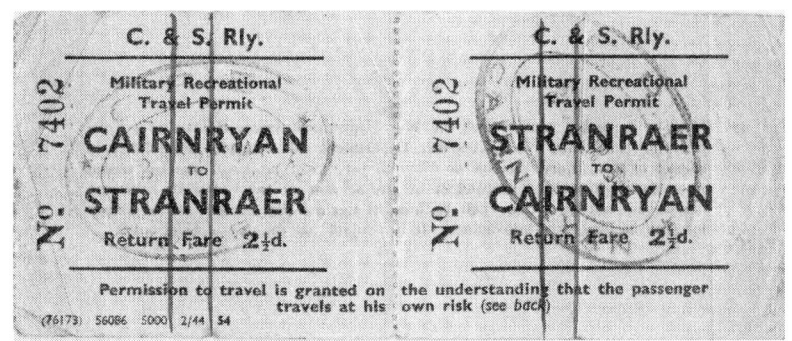

A rare souvenir of the Cairnryan Military Railway, a military recreational travel permit, issued in March 1944. Note that whoever was responsible for ordering these tickets from the printer thought of it as the Cairnryan & Stranraer Railway.
*Nelson-MacDonald Collection*

YMCA Halt does not appear to have been built. The average journey time was twenty-two minutes, not bad going for six miles with two halts. First and third class coaches were provided and the return fare, all the way, was two and a half pence. It must be realised however that passenger travel had a low priority: freight was what mattered.

Mrs Ena Scott from Drummore in a letter to the *Wigtownshire Free Press* recalled that in 1945 as an ATS driver in Cairnryan, she "... bought on payday a railway ticket for two and a half pence which allowed her to travel on the train any time in that week ... the seats were wooden and there were no window panes, no lights, no heating .... nevertheless no-one complained." Again, according to David L. Smith, the coaches were non-corridor, and built in 1911!

In June 1942, before the official opening, King George VI and Queen Elizabeth, the late Queen Mother, travelled on the Cairnryan Military Railway (CMR) from Cairnryan Junction to

Claddyburn Halt, looking north, with Cairnnryan Manse on the right, September 1944.

Courtesey of the Imperial War Museum

The engine shed at Leffnoll with locomotive *The Princess Royal* on the left, September 1944.  *Courtesy of the Imperial War Museum*

Cairnryan to board the cruiser HMS *Phoebe* for their visit to Northern Ireland. The Royal Train consisted of two coaches specially fetched in for the occasion. Their Majesties returned after a few days. Security was very tight. The driver on both occasions was a Stranraer man, George Harvey.

Charles Mcacher has a good story about HMS *Phoebe's* return. Once Their Majesties were safely on their way, shore leave was granted. Six hundred sailors swarmed ashore demanding to be directed to the village inn. An unsympathetic Sapper informed them that they were much more than a few years too late, and, he added as an afterthought, that the last train for Stranraer had just left.

There is a tradition that the then Princess Royal on one occasion drove a locomotive part of the way on the line. The locomotive was then and there named *The Princess Royal*. Certainly an Imperial War Museum photograph (H40128) gives one of a trio of locomotives in the Leffnoll engine shed the Royal title.

One aspect of the railway construction was the enormous number of workers needed. At the peak, 4,000 servicemen were employed on the railway and port building. The problem of housing that number and providing for their welfare brought about a mushroom-like eruption of military camps, named mostly after the farms on whose land they were built. From the beginning of 1941, units of Royal Engineers and Pioneer Corps had been involved in camp construction.

Apart from Transit Camp, on the outskirts of Stranraer in Cairnryan Road, providing temporary accommodation for troops going to or returning from Northern Ireland, many other sites were built on. Aird Camp catered for the labour force at Exchange Sidings, Rock McGibbon (Innermessan) for the RAOC and REME in the repair yard. The massive Leffnoll Camp served the marshalling yard. Drummuckloch had a large MIT contingent, Bankhead and Meadowpark probably the smallest. Claddyburn on the north of the stream of that name held the

Crest left by men of the Pioneer Corps to mark their work on the railway. *Courtesy of Stranraer Museum, the Bill Gill collection*

personnel from the Lighterage and School Sidings while the Officers' Mess above the village was late built and tradition again says little used. A large camp, Cairnryan, above Lochryan House was home to servicemen and women of the Military Port and Administration Block. Lastly, Bonnybraes Camp at the north end of the village was linked with the Pile Construction Yard. Accommodation in all cases included office, guardroom, cookhouse, dining hall, ablution and latrine blocks, and canteen.

The Port was opened officially in July, 1943 though shipping had started to arrive and depart some months before. In February, 1944 the centre of activities moved to the South of

The administration block for Cairnryan Camp and Port, September 1944.

*Courtesy of the Imperial War Museum*

This photograph, taken in September 1945, shows personnel preparing one of the fleet landing craft which were used for dumping deadly cargoes of shells, bombs and other surplus explosives from Cairnryan after the Second World War.

*Courtesy of the Imperial War Museum*

England in preparation for D-Day. Cairnryan Port was then put on a care and maintenance basis. It had never worked to full capacity. The railway continued operating as the occasional ship berthed. From 1943, the "Milk Run" from Larne was switched from Stranraer to Cairnryan. The *Whitstable*, loaded with metal milk containers, arrived at the South Deep at 5 a.m. and the liquid cargo was manhandled on to the waiting rail wagons before setting off for Cairnryan Junction and destination Glasgow. After an unsuccessful experiment to carry the milk by air transport, sailings to the Cairn were resumed in November, 1948, when civilian labour was substituted for military. In 1949, when the car deck of the *Princess Victoria* was strengthened, road tankers with their contents could be carried to Stranraer.

The Port may have been put on a care and maintenance basis in February, 1944 but not so the Pile Construction Yard where construction work for the Mulberry Harbour saw the workforce there working flat out for the next four months. With the coming of D-Day in June, work ceased and numbers were reduced drastically. In 1946, Harry McKenzie, on National Service in the Royal Corps of Signals, remembers the place as deserted except for a watchman, an elderly Sapper, living all alone in one of the huts, who only emerged for pay parade and to draw rations.

Yet at the War's end when it looked as if the Port and the Railway had fulfilled their purposes, activity descended on them once more. The Government found itself holding vast quantities of explosives unneeded in peacetime. The quickest and cheapest method of disposal was dumping at sea, and an underused port with its own railway and military operators was just what was needed. Over the next few years the railway was to carry countless wagonloads of explosives of all kinds. Bombs, large and small, mines, sea and land shells, grenades, small arms ammunition; and fuses poured into Cairnryan. Gas shells were loaded on to unwanted ships, like the *Empire Claire*, which sailed on a last voyage to be scuttled at sea. A particularly horrible cargo of German nerve gas arrived by sea from Wales and was likewise committed to the deep. The village was not informed though the Education Committee of the County Council bought the YMCA Canteen and evacuated the village school there.

David L. Smith in *The Little Railways of South-West Scotland* relates a happening which might have had a horrific effect on the south west and could well have happened on the CMR. Two ammunition trains loaded with gas shells were passing through Girvan on a very dark, stormy night bound for Cairnryan. The crew of the second train standing at the station were petrified when out of the darkness a runaway group of wagons from the first train hurtled towards them. This might have been the rail disaster of all times had not the signalman at Girvan switched the points over just before the wagons broke loose. A narrow escape for Girvan.

Ordinary shells were loaded in their boxes onto a fleet of landing craft, taken out to sea, unpacked, and rolled down chutes to the water. Although all ammunition was supposed to be defused before arrival, tragedies could happen. Eight young servicemen, members of an RE section loading shells on the North Deep Water Wharf, were blown to pieces when a case of fuses was accidentally mishandled. Four of them are buried in Stranraer's Glebe Cemetery, three "known only to God". A very detailed account of this traffic is given by Richard Holme in his *Cairnryan Military Port*, published in 1997.

In 1959 military activity had almost ceased, the workforce being largely civilian. The War Department withdrew all service personnel, called it a day, and closed the Port. Only a skeleton train freight service was left to operate for the new owners.

The signal box at Cairnryan Junction remained open for a time just in case a use could be found for the railway. (Incidentally, that signal box was one of the few locally that used female operators.) Gradually the Nissen huts were sold off and disappeared. The final locomotive used for lifting the rails, for sale abroad, hooted its last in 1967. A line that had taken eighteen months to construct finally vanished in a few weeks.

A photograph taken from the 60-ton crane in 1944 shows how the lighthouse, which once occupied Cairn Point on its own, was flanked by the North and South Deeps.

*Courtesy of the Imperial War Museum*

View of Cairnryan harbour from the south in the early 1950s. *Oakwood Press collection*

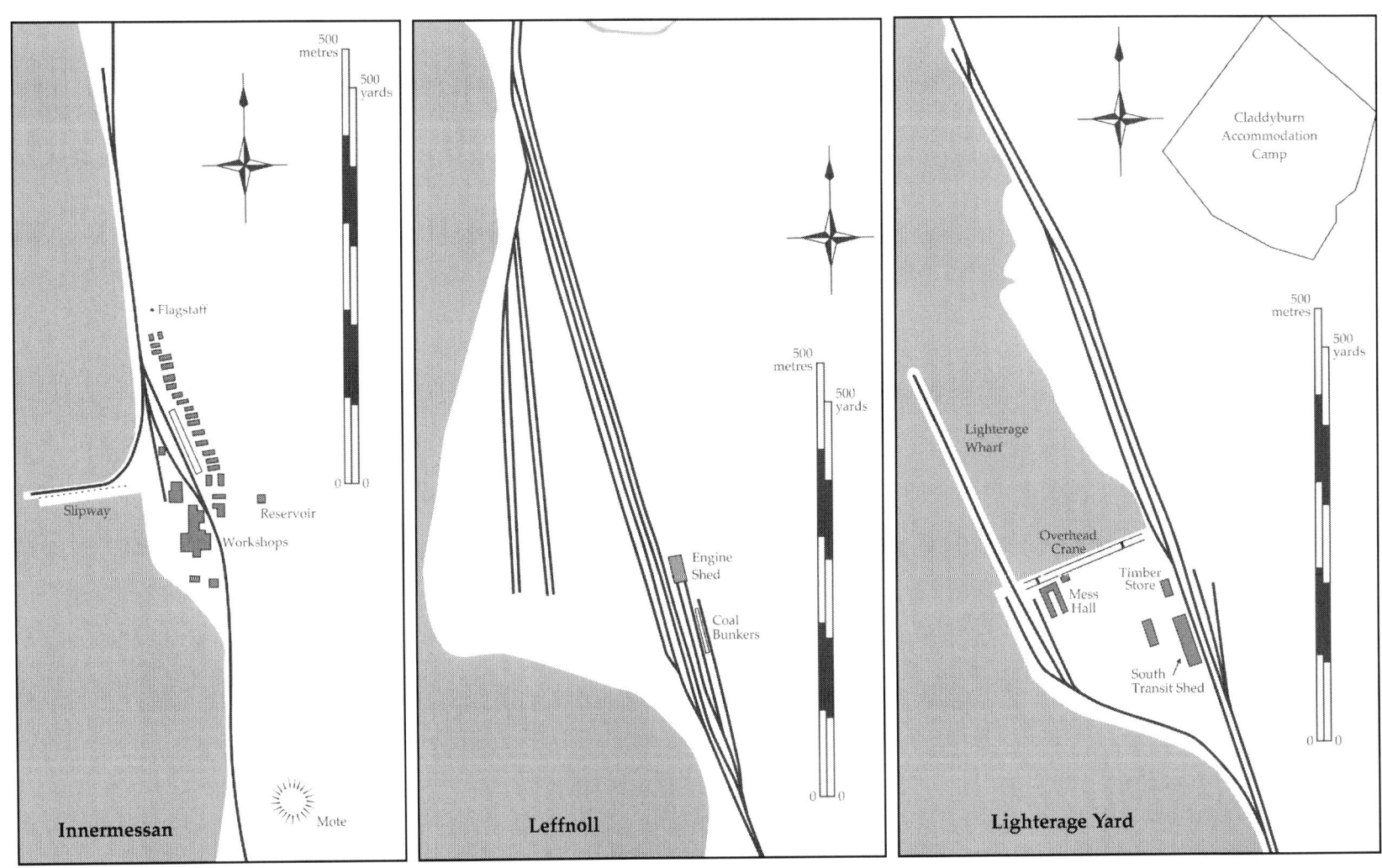

Track diagrams for the line redrawn from the 1957 six inch to the mile OS map held by the National Library of Scotland. The line was probably surveyed in 1951 when the sheet was revised. By that time there had been some track removal, Leffnoll Marshalling Yard is a shadow of its wartime height.

North
Deep Water
Wharf
Bonny Braes
Camp
Stores and
Offices
Church
North
Transit Shed
Lochryan
House
Cairnryan
Camp
Administration
Offices
Lochryan
Lighthouse
Fire
Station
Office
Office
500
metres
500
yards
0
0
Officer's
Mess
Stores and
Offices
Cairnryan
South
Deep Water
Wharf
Cairnryan Port

Cement
Store
Offices
Casting
Beds
Workshop
Stores and
Mess
Oyster
Company's
Sheds
Old House Point
North
Deep Water
Wharf
Bonny Braes
Camp

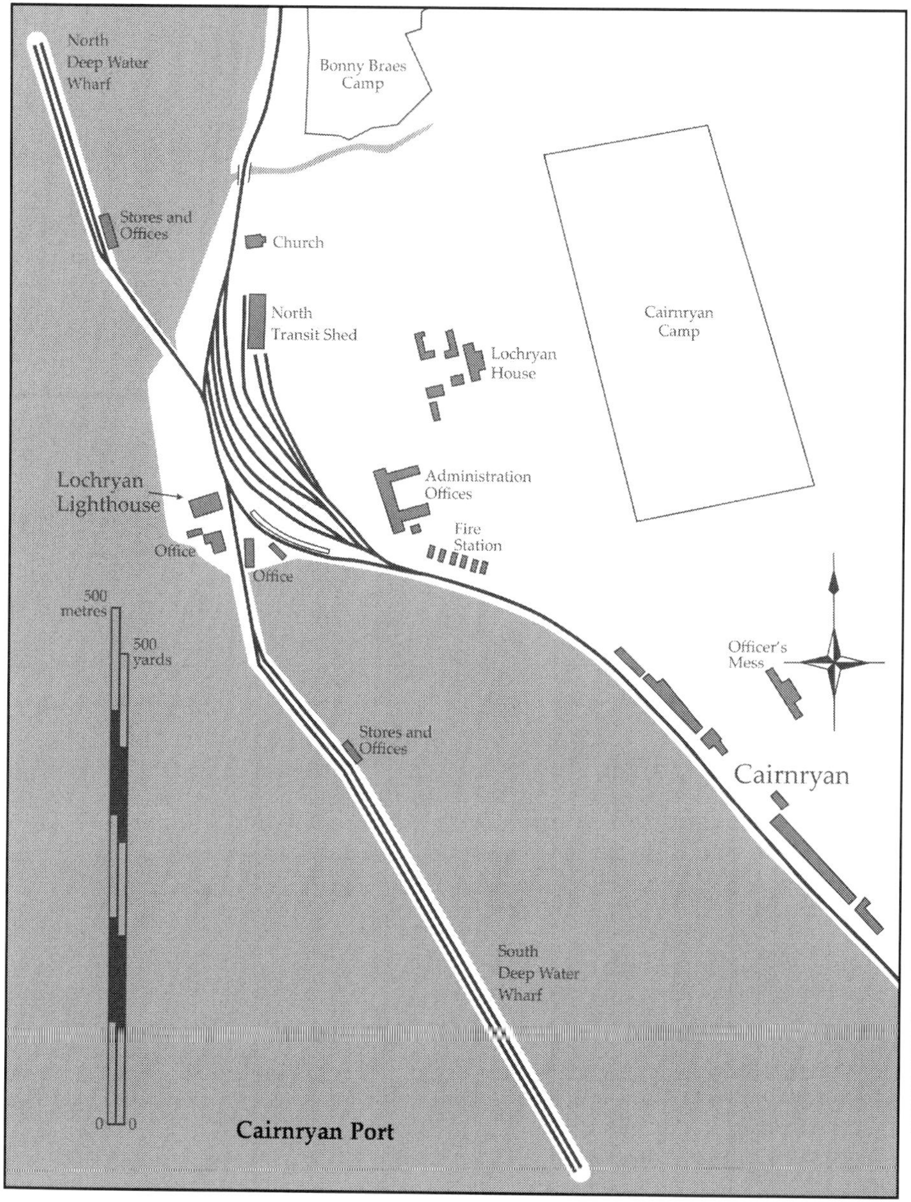

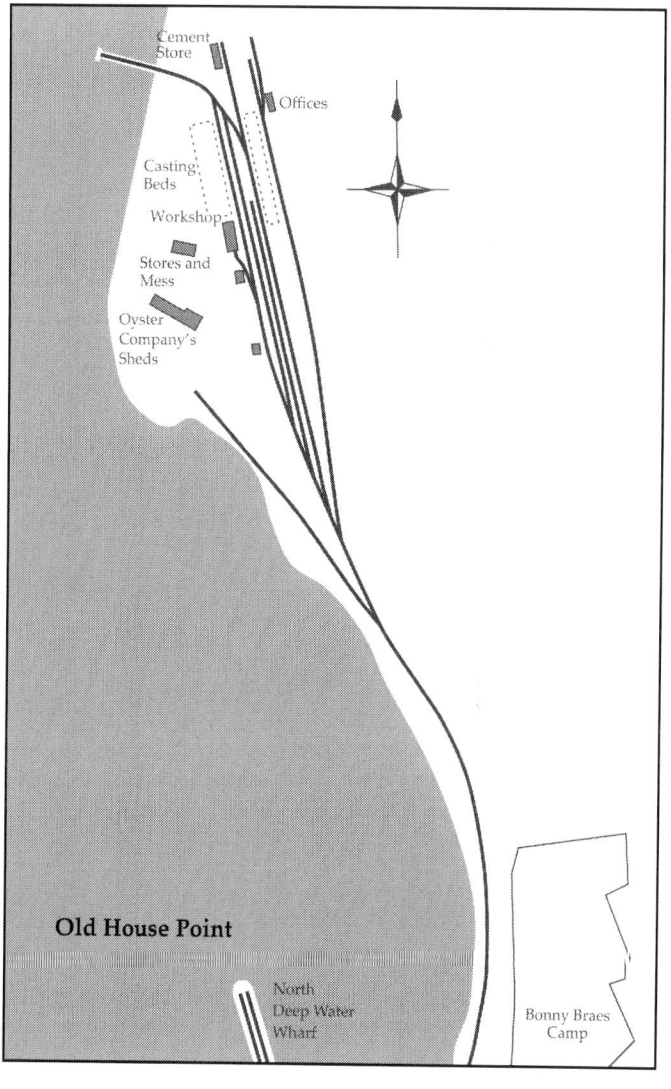

Innermessan Camp, and workshops, looking north, September 1944. Many of the workshop buildings pre-dated the building of the railway and were built in 1917 by the Multilocular Shipbuilding Company. Their intent was to build concrete ships at the site, for the Merchant Navy as replacements for ships lost during the war. The *Liverpool Journal of Commerce* wrote a description of the site on 21st January 1919. They described the workshops and processes of the company, and the advantages of the modular concrete building method over monolithic casting of ships.

*Courtesey of the Imperial War Museum*

Innermessan Camp, workshops and pier looking west, September 1944. The company clearly struggled. In 1920 they advertised concrete parts for making culverts, then in September that year they advertised their concrete working tools for sale. For the rest of the 1920s they focused on shipbreaking. In 1921 two employes of the company drowned, when the small boat they were using was dragged underwater by the weight of the steel hawser they were taking to *HMS Pactolus* which was being broken up. The end of the company came in the early 1930s when it was liquidated.

*Courtesey of the Imperial War Museum*

# THE CAIRNRYAN MILITARY RAILWAY TODAY [1999]

It is still possible to trace the route of the railway all the way from Cairnryan Junction to the Old House Point. Unfortunately it is no longer possible to walk it all the way. Like any other disused line in the region there are parts where bridges have been demolished or which are now private ground. Many places have been surrendered to the wild and have become impenetrable through rampant growth of whins (gorse) and brambles and young forest trees. On one stretch, tidal erosion makes progress difficult. However, it presents a worthwhile challenge. Walking is pleasant for much of the way and even where there are obstructions a reasonably fit person should be able to cope.

At the beginning, Cairnryan Junction, the actual spot where the line branched off from the main line, is difficult to find despite being only a matter of yards from the modern bridge north west of Limekiln Kennels in Limekiln Road, (map reference 087599). A car can be pulled off the road on either side of the bridge, where on the left the low embankment is fairly visible. Some little distance along, the huts of the Yard Foreman and MC Detachment still stand and are in private use. Exchange Sidings, marked by different heights of grading, are now encountered and the going is rough. It is of course illegal to trespass on the railway line and if progression, bent almost double at times beneath a natural plantation of young trees, becomes too much to bear, then cheat a little. Return to the car and park it about three quarters of a mile along the road nearly opposite the unusually designed house on the right.

Follow the track through part of what was once Aird Camp by concrete foundations, and an old wire fence gives access to Cairnryan Railway. Progress is not too difficult but slow. Shortly the walls of Aird Block Post can be seen on top of a low

The water tower, capped by ivy, and coaling stage (left) at Aird, September 2021.
*Oakwood Press*

embankment, fairly well preserved but difficult to inspect. Less than two hundred yards away the tall water tower rears above the undergrowth. This also seems to be in a fair state of preservation, now standing on a brick foundation. Close by the coaling stage is found but again is difficult to access.

This was a busy area. The CMR curves past the coaling stage and soon runs high on the long embankment dropping gradually to Bridge of Aird. The new road to the industrial estate cuts through the near end of the embankment. The route of the temporary line of 1942 can be followed to that road, where it continues down through a plantation and then runs down the east side of Aird Donald Caravan Park. The owner, Bert Cassie, shows proudly the remains of the huge Command Post that co-ordinated the air defence of the Loch Ryan area. The concrete foundations on which many of the caravans are sited once supported Nissen huts occupied by prisoners of war. The railway bridge over the Black Stank, probably Bridge No. 1, has long since disappeared. After this no traces of the temporary line are visible. Back at the embankment follow the road to the junction with Limekiln Road, where the brickwork of Bridge No. 2 can be seen, then follow the road to the junction with the A 75, where the brickwork of Bridge No. 4 is evident on the far side. The girders of both bridges were removed a good few years ago. Turn left on the A 75 and in a hundred yards turn left again past Bridge of Aird Farmhouse and cross Old Bridge of Aird. Right in front the other side of Bridge No. 2 can be seen and to the left the distinctive culvert under the embankment, Bridge No. 3, takes the Black Stank through to New Bridge of Aird on the A 75.

The embankment itself is well overgrown but ponder on the effort and labour that went into its construction. The line curves to the right and traces of Construction Junction are hard to find. The remains of Construction Junction Block Post are obvious though. Here the line ran between the Bishop Burn and the A77 and crossed that stream by Bridge No. 5. The girders are still in place though no longer parallel. They are easily crossed by those with a good sense of balance. For those without such a sense, the stream is easily fordable.

Just visible here, the line next crossed the minor road from Sandmill to Inchparks Croft to run on the flat, over two level crossings on the farm roads to Low and High Balyett, from which it curves left to meet the A77 by the crossing keeper's hut, still standing. A line of chalets and static caravans in Innermessan Caravan Park marks its progress through that complex. Approach, however, should be by the main entrance. Girders still in position mark the railway bridge over Messan Burn, alongside the old turnpike road bridge. Erosion has made the next stretch difficult but not impossible and once the barrier of huge boulders (anti-litter dumping) is reached, Innermessan or Rock McGibbon Camp can be seen and explored. A few brick huts still stand along with a somewhat ornate lavatory and enough concrete foundations to help mentally reconstruct the

Remains of Bridge No.4 and the embankment leading towards Loch Ryan, September 2021. *Oakwood Press*

Remains of slipway and pier at Innermessan, January 2015.

*Billy McCrorie*

Old block house at the Beoch Burn, January 2015.

*Billy McCrorie*

scene of 1941-59. The jetty has been badly treated by the storms as has the main slipway though a smaller one is more or less intact. The small wreck north of the jetty, visible at low tide, is not as old as the camp. Rock McGibbon itself is also seen at low tide. The huge concrete sections lined up as if on parade could have been towed there from the Pile Construction Yard possibly having been surplus to requirement. The track leads on the left of the roadway, protected by boulders, some of them sited very recently. Another anti-litter barrier must be squeezed through as the line leaves the road where a railwayman's hut is obvious. A few cars can be parked where the road joins the A77.

The embankment stretches away to the north, following the HWM, protected by a concrete casing, now crumbling in places. The Pioneer Corps badge mentioned earlier will be found near the commencement of the concrete though not so visible as it was even ten years ago. The track leaves the shoreline after three hundred yards. Do not follow it. It disappears into a jungle of brambles that camouflage its crossing of the Beoch Burn by a simple girder bridge, still in place but somewhat untrustworthy. It is visible from the road above.

Keep to the shore and round the point to the estuary of the Beoch Burn. At low tide it is fordable but at other times, as previously mentioned, "wellies" are useful. They have their use also for negotiating low brambles and whins, while walking on the long stretches of springy turf is easy and pleasant. Many dog owners exercise their pets here. Follow a track to the right and soon you will come to the A77. Just before the exit, on the left, stood the engine shed. Absolutely nothing remains. Right about turn and the coaling stage is in view more or less complete. As seen on the previous coaling stage at Aird, concrete and brick do not invite demolition. Behind the coaling stage can be found Leffnoll South Signal Box, on the odd occasion used by a travelling person as a doss house.

Now retrace your steps to the shore with a word of warning. There are many interesting paths winding their way through the clumps of whins and tangles of brambles. Do not attempt to follow them on your own. That sort of area is seldom visited and a sprained ankle could keep a person calling for help for an hour or so before someone comes along, then the caller may take some finding. By keeping to the shore, good progress will be made but some more of the remaining buildings will be missed, e.g. the Yardmaster's Office, the RTO Office and the Leffnoll North Signal Box. These can be seen and even visited from the A77 as can the small girder bridge over the Leffnoll Burn, only a short walk from the Rhins of Galloway Hotel.

As you continue along the shore the turf gives way to whins overgrowing a bank protected by rubble and a concrete shoe, which makes for easy walking. Towards the end of the concrete a few pebble inscriptions can be found, some illegible but one reads "50th PC 1943", commemorating repairs effected after the storms of winter, 1943. Once round the point the

The remains of the bridge over the Beoch Burn, January 2015.
*Billy McCrorie*

Coaling stage at Leffnoll, burried among the brambles, January 2015.

*Billy McCrorie*

The remains of the girder bridge over the Several Burn, September 2021.

*Oakwood Press*

Several Burn is easily forded at low tide. A substantial embankment is now apparent, well shored up with large boulders. The ubiquitous brambles have spread thickly along the top. A concrete buffer stop will be met, and a few steps farther on a break in the overgrowth gives a view of the Rhins of Galloway Hotel, built in 1943 as a YMCA. The thick brambles cling to the site of the YMCA Halt platform, a wooden contraption which has left absolutely no trace behind. After the Education Committee closed Cairnryan School, the building, like so many of its kind, became a tearoom, then a hotel. From the point above a splendid view of the shoreline, the railway track and the A77 road can be had. While the huge camps of Leffnoll and Drummuckloch can still be recognised from their concrete foundations, the scar of the quarry below Bankhead has healed with time.

After 1959 the Lighterage was bought from the War Department by Messrs Pounds of Portsmouth along with the rest of the Military Port and the northern half of the railway. The Atlantic Steam Navigation Company bought the Lighterage in 1964. It was developed as a ferry terminal in the early 70s and then bought by European Ferries (Townsend Thoresen) in 1973. P&O acquired it in 1987. Improvements have swept away almost all of the military remains. Rails, cranes, and the Transit Shed have gone and School Sidings has become a car park.

The pond or "Loop", where the timber was stored in 1941, can be visited but the double track to the Cairn Point is now a designated walkway, permitting the village to be seen as it stood before 1941. There are two lay-bys along its length for viewing the Port. The six-feet-high wire fence is now thankfully a thing of the past. Tired of complaining about a structure that no longer served a security purpose and parts of which were apt to blow down across the road in a gale, the Kirk Session sallied out one evening from the Church Hall and demolished a large part of it. Across from the Lighterage and behind Lochryanhall, now Lochryan Hotel, Borrow Pit, which had given much of its

The South Deep Wharf from Cairnryan, the structure is now in poor repair and there are several place where collapses have narrowed it, August 2018. *Billy McCrorie*

Looking over the security fence at Cairnryan to the lighthouse. Between it and the other building, originally an electrical substation, is a railway wagon, September 2021. *Oakwood Press*

contents to construct the Lighterage, became first a vehicle park then in the late 1950s a Pioneer Corps camp when Claddyburn Camp, built only for the duration, was no longer habitable. Its foundations are still visible as foundations for the chalets in Cairnryan Caravan Park.

Towards the end of the village, the ruins of the Officers' Mess look down on the South Deep Water Wharf, carrying a huge load of quarry material and at the time of writing out of use. To the north of the Mess and above Lochryan House the remains of Cairnryan Camp are gradually being covered by trees. From it Bonnybraes Camp is seen to be in a similar condition. However, all traces of the once distinguished administration building have disappeared completely with the exception of a derelict air raid shelter. Another car park has taken its place. The telephone exchange and fire station have also gone.

Rubble Bank Sidings have been tidied up since the days of shipbreaking, which saw the end of the great aircraft carriers *Centaur*, *Eagle*, and *Ark Royal*. A few original buildings still stand as stores and offices and watchman's quarters, now owned by Messrs Barr Construction. The Block Post at the port entrance is still there as is the rail platform, but the North Deep Water Wharf was demolished in 1960. The lighthouse now shows a red beam to avoid confusing the ferry traffic. At the north end of the Sidings the North Police Hut still guards the entrance over Glen Burn but the bridging girders have gone. The single track has been converted into an attractive picnic area.

It stops just short of where the line ran into Old House Point, where an enclosure on the right hides the burial place of quantities of asbestos from the various ships broken up. Immediately visible are the ruins of the once habitable cottage and the large workshop beyond it. A cookhouse chimney stands on its own in front of the cottage and the remains of a ramp for tipping gravel will be seen on the shoreline. What is left of the once flourishing oyster tanks is nearhand. The tanks are still intact but the superstructure surrendered to a vicious gale some years ago. Concrete supports for Mulberry Harbour construction stand up in profusion and the swell of the ferries washes over the few rejected concrete barges or "Beetles" which provide platforms for sea anglers. Three or four rotting timber piles remain of the jetty where Charles Meacher tended his steam engine.

But still intact, on the east side of the Pile Construction Yard buried again under a pall of whins and visible only if a passageway can be forced through, are some two hundred yards of casting beds. At the south end, a huge stockpile of deck beams, weighing it is said some ten tons each, awaits a ghostly train of wagons of the Cairnryan Military Railway.

Finally, it cannot be emphasised enough that most of the sites mentioned are on private ground. Permission to inspect and to park where necessary should be sought.

The concrete abutments are all that remain of the bridge over the Glen Burn, the notches being where the girders sat, February 2014.
*Billy McCrorie*

Former officers' accommodation at Cairnryan, July 2012.

*Nelson-MacDonald Collection*

The end of the line. In The 1960s the rail lines which had served Cairnryan Port were lifted and sold. The last of the little engines which carried out this work was captured on film by the author. *Courtesy of Stranraer Museum. The Bill Gill collection*

No. 3193 *Norfolk Regiment* is a Hunslet "Austerity" tank engine built in Leeds in 1944 for the War Department. She ended her army service in 1952-53 on the Cairnryan Military Railway where she was used to haul trains of spent and captured munitions for dumping in the sea. Thereafter she had a hard life with the National Coal Board (she was once knocked over on her side by a runaway train of coke wagons) before being bought into preservation in the 1970s. She was acquired in 2009 by a charitable trust in Norfolk, requiring extensive boiler work and arriving as a kit of parts on four lorries. She was re-assembled and restored over the next ten years, given a name to fit her military heritage, and is now based at Bressingham Steam Museum in Norfolk but goes out on hire to pull passengers on various heritage railways. She is painted black as it is believed she was at Cairnryan, but the War Department lettering is largely guesswork, and the army certainly wouldn't have given her those lovely red coupling rods! For such an old lady she has proved tough, reliable, rather slow but very powerful and much-loved by those who drive her.

*Norfolk Heritage Steam Railway Ltd*

Photograph taken from the 60-ton crane on Douth Deep Water Wharf, showing the North Deep on the left. The buildings at the landward end of the pier were stores and offices. Behind them in its camouflage paint is the North Transit shed. The huts of Bonnybrae Camp are on the hill on the right.

*Courtesy of the Imperial War Museum*

# Appendix

One of the few original contributions dealing with the day-to-day work on the CMR and the Military Port is the painstaking research carried out by Major J. Starling RPC on the war diaries of the various Pioneer Corps companies stationed in and around Cairnryan and Stranraer from 1940 to 1959. Some of the diaries were kept meticulously, e.g. 13 Company 1951-59, some leave much to the imagination, and others owing to circumstances and hurried moves convey very little information. There are no records for the period 1945-48.

An abridged version of the research follows:

**168 Company**            Oct 40 - Jan 41
02 Oct 40   H.Q. and Half Company move to Stranraer from Barry - camp construction.

**152 Company**            Nov 40 - Mar 41
30 Nov 40   Detachments in Stranraer employed in hut erection.

**71 Company**            Jan 41 - Oct 41
6 Jan 41   H.Q. and Half Company move to Stranraer – hut erection, road work, and pipe laying.

**257 Company**            May 41 - Mar 44
15 May 41   H.Q. and Half Company move to Stranraer for dock construction.
Aug 41   Whole Company now in Stranraer for dock and camp construction.
01 July 43   Whole Company in port construction.
31 Dec 43   Company on camp construction less one Section on port construction.

**267 Company**            Dec 41- Oct 43
31 Dec 41   4 Sections are working at Glenluce with the Dept. of Agriculture.
Moved to Stranraer for port and camp construction.
30 Jun 42   Dock construction at Loch Ryan
31 Dec 42   5 Sections on railway construction.
2 Sections on camp construction.
3 Sections on quarter construction.
1 Section on cable laying.
2 Sections on port construction.
1 Section at Central Supply Depot.
1 Section on dam construction.
30 Apr 43   13 Sections on port construction.
1 Section on reservoir construction.
31 Jul 43   10 Sections on port construction.

**196 Company**            Oct 41 - Dec 44
14 Feb 42   Port and camp construction.
18 Aug 42   Port and camp construction.
3 Dec 43   Company still operating No. 2 Military Port.

**272 Company**            Apr 42 - Jul 43
28 Apr 42   Arrived Stranraer.
28 Feb 43   Port construction and sea wall construction.

**50 Company**    Aug 42 - Jul 43.   August 43 - Sep 43
23 May 42   Arrive Lochans.
26 Aug 42   Moved to Stranraer.
Dec 42   Camp construction, cable laying, dam construction, and storage tank construction.

**137 Company**            Dec 43 - Feb 44
13 Dec 43   Loading and unloading ships.

Early 1950s view of the South Deep Water Wharf. After the Second World War, the port fell into disuse, and the wharfs and cranes were idle. The cranes were sold to the Morrocan port of Agadir to help with rebuilding its facilities after the 1960 earthquake. Cairnryan would continue to decay and became a shipbreaking yard in the early 1960s. On 10th July 1973 the European Ferries Group started sailing from Cairnryan to Larne with *MV Ionic Ferry* from the old Lighterage wharf.

*Nelson-MacDonald Collection*

**12 Company formed Feb 49 from 13 and 71 Companies Located at Cairnryan.**

17 Oct 49    Pte. Thomas Daggett killed loading ammunition. Camp shared with 1 EDU RAOC. Disbanded by Sep 56.

### 13 Company      1952 - Apr 59

2 Feb    Entire Company moved to Loch Ryan for ammunition dumping.
Mar 51    Half Company now serving in Cairnryan on sea dumping and rail maintenance.
Dec 51    Loaned a canteen by the Church of Scotland.
Dec 52    Support for C Company No. 1 DUR RAOC Wig Bay.
Jun 55    NAAFI has been built at Quarry Camp.
7 Sep 56    2324176 Pte. P. McGuire drowned.
Sep 57    Dumping in Atlantic, 200 miles away, from LST *Empire Curlew*. Each journey takes four days and occurs every two weeks.
Dec 57    New married quarters ready.
Mar 58    LCTs arriving for "summer cruises" to St Kilda, a 3 to 4 week journey on resupply.
Jun 58    Operation "Hardrock" in the support of St Kilda. A ten-day job is undertaken to dump ammo from Milford Haven which is unsafe to move by rail.
Sep 58    A Detachment of 1 officer and 20 men to St Kilda for 1 month - preparation of base. Port closes for two weeks.
Dec 58    Cairnryan Military Port due to close Apr 59. St Kilda run stops - winter. Visit from HMS *Britannia*, two submarines, and an ocean cable layer.
Mar 59    The Navy discovers some thousands of depth charges for an urgent "dump".
Apr 59    Disbands. (The unit had remained in Quarry Camp from Mar 51).

### 421 Company      Aug 47

Aug 47    Arrived Stranraer. Located in Cairnryan as part of I 03 Group. Deep sea dumping.
Dec 47    HMV *Loighton* containing 1000 tons of mustard and Lewisite gas is sunk. Disbanded N.K.

Ammunition dumping barge at South Deep.
*Nelson-MacDonald Collection*

On the opposite side of Loch Ryan from Cairnryan was Wig Bay. It saw some use during the First Word War for seaplane testing, a role that was expanded during the Second World War when the bay became a servicing area for flying boats. There were two slipways and plenty of mooring space in the loch. In June 1940 it became the site of the Floatplane Training Flight and the Flying Boat Training Squadron which moved there from Calshot, on the Solent, due to the German occupation of France. 240, 209 and 228 Squadrons of Coastal Command were based at Wig Bay and carried out patrols over the Atlantic supporting British convoys. Storage of aircraft began during the Second World War with the formation of No. 57 Maintainance Unit, and increasing numbers of flying boats were stored and maintained there. After the war the work changed to disposing of the aircraft, No. 57 was disbanded in 1951, with a team from Short Bros and the RAF completing the job in 1957.

*Nelson-MacDonald Collection*